SCIENTIFIC AMERICAN EDUCATIONAL PUBLISHING

SCIENTIFIC AMERICAN INVESTIGATES THE HUMAN BODY

THE SKELETAL SYSTEM

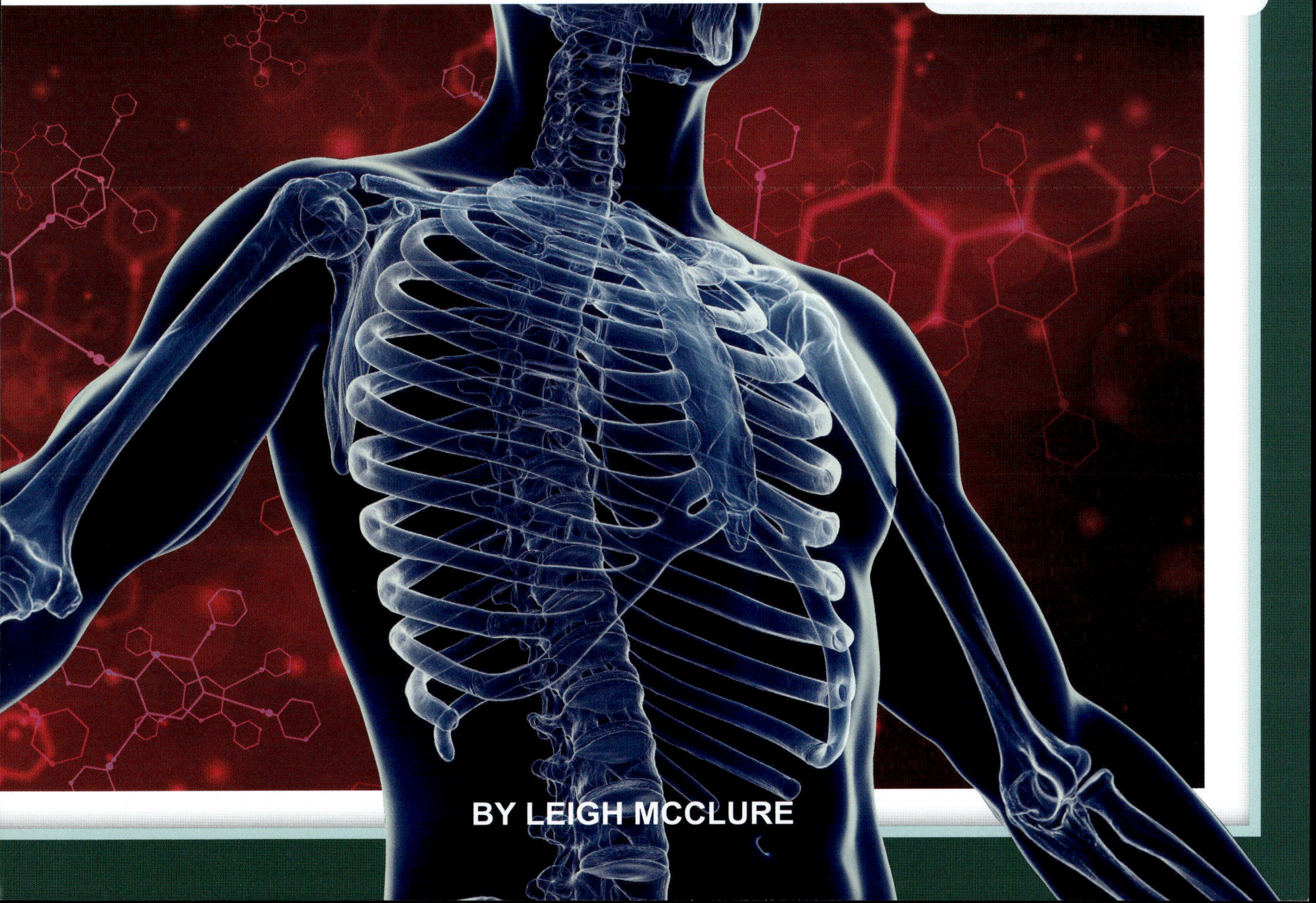

BY LEIGH MCCLURE

Published in 2025 by The Rosen Publishing Group
in association with Scientific American Educational Publishing
2544 Clinton Street, Buffalo NY 14224

Library of Congress Cataloging-in-Publication Data
Library of Congress Cataloging-in-Publication Data Names: McClure, Leigh, author. Title: The skeletal system / Leigh McClure. Description: Buffalo, NY : Scientific American Educational Publishing, [2025] | Series: Scientific American investigates the human body | Includes bibliographical references and index. Identifiers: LCCN 2024011263 (print) | LCCN 2024011264 (ebook) | ISBN 9781725351516 (library binding) | ISBN 9781725351509 (paperback) | ISBN 9781725351523 (ebook) Subjects: LCSH: Human skeleton–Juvenile literature. | Bones–Juvenile literature. | Musculoskeletal system–Juvenile literature. | Human anatomy–Juvenile literature. Classification: LCC QM101 .M328 2025 (print) | LCC QM101 (ebook) | DDC 612.7/51–dc23/eng/20240326 LC record available at https://lccn.loc.gov/2024011263 LC ebook record available at https://lccn.loc.gov/2024011264

Designer: Andrea Davison-Bartolotta
Editor: Caitie McAneney

Portions of this work were originally authored by Greg Roza and published as *The Skeletal System*. All new material in this edition authored by Leigh McClure.

Photo credits: Cover, p. 1 (main) SciePro/Shutterstock.com; cover, p. 1 (background) jijomathaidesigners/Shutterstock.com; p. 5 (main) CLIPAREA I Custom media/Shutterstock.com; p. 5 (inset) Jacek Chabraszewski/Shutterstock.com; p. 7 (inset) graphicgeoff/Shutterstock.com; p. 7 (main) New Africa/Shutterstock.com; p. 9 elenabsl/Shutterstock.com; p. 11 (inset) Designua/Shutterstock.com; p. 11 (main) dotshock/Shutterstock.com; p. 13 BearFotos/Shutterstock.com; p. 15 Pixel-Shot/Shutterstock.com; p. 17 Xray Computer/Shutterstock.com; p. 19 (main) Terelyuk/Shutterstock.com; p. 19 (inset) vetpathologist/Shutterstock.com; p. 21 Roman Chazov/Shutterstock.com; p. 23 (main) Midkhat Izmaylov/Shutterstock.com; p. 23 (inset) marina_ua/Shutterstock.com; p. 25 mr.suphachai praserdumrongchai/iStockphoto.com; p. 27 (main) Prostock-studio/Shutterstock.com; p. 27 (inset) art4stock/Shutterstock.com; p. 29 Sergey Novikov/Shutterstock.com.

Printed in the United States of America

CPSIA compliance information: Batch #CSSA25. For Further Information contact Rosen Publishing at 1-800-237-9932.

CONTENTS

YOUR SUPER SKELETON 4

ALL ABOUT BONES 6

WHAT ARE BONES MADE OF? 10

JOINTS AND MOVEMENT 12

YOUR BODY'S SUPPORT SYSTEM 14

BONES AS PROTECTION 16

CELL PRODUCTION 18

SAME, BUT DIFFERENT 22

BREAKS AND BRUISES 24

BONE DISEASES AND DISORDERS 26

KEEP YOUR BONES HEALTHY! 28

GLOSSARY 30

FOR MORE INFORMATION 31

INDEX 32

Words in the glossary appear in **bold** type the first time they are used in the text.

YOUR SUPER SKELETON

You're probably familiar with the image of a skeleton. Each Halloween, they are displayed as spooky decorations and symbols of death. However, your skeleton is more than just a decoration. And you couldn't live without your bones!

Bones give structure to your body. Without them, you'd just be a blob of skin and organs. They help you produce important cells, move around, and even hear with your ears.

The largest bone in the body is the femur, which is your thigh bone. For adults, the femur is around 18 inches (45.7 cm) long. The smallest bone of the body is in the middle ear, and it's only a few millimeters long! From big bones to small, the skeletal system plays an important part in the activities you do every day.

FUN FACT

BABIES HAVE ABOUT 300 BONES. HOWEVER, AS PEOPLE GROW, SOME OF THEM FUSE, OR GROW TOGETHER. AN ADULT BODY USUALLY CONTAINS 206 BONES.

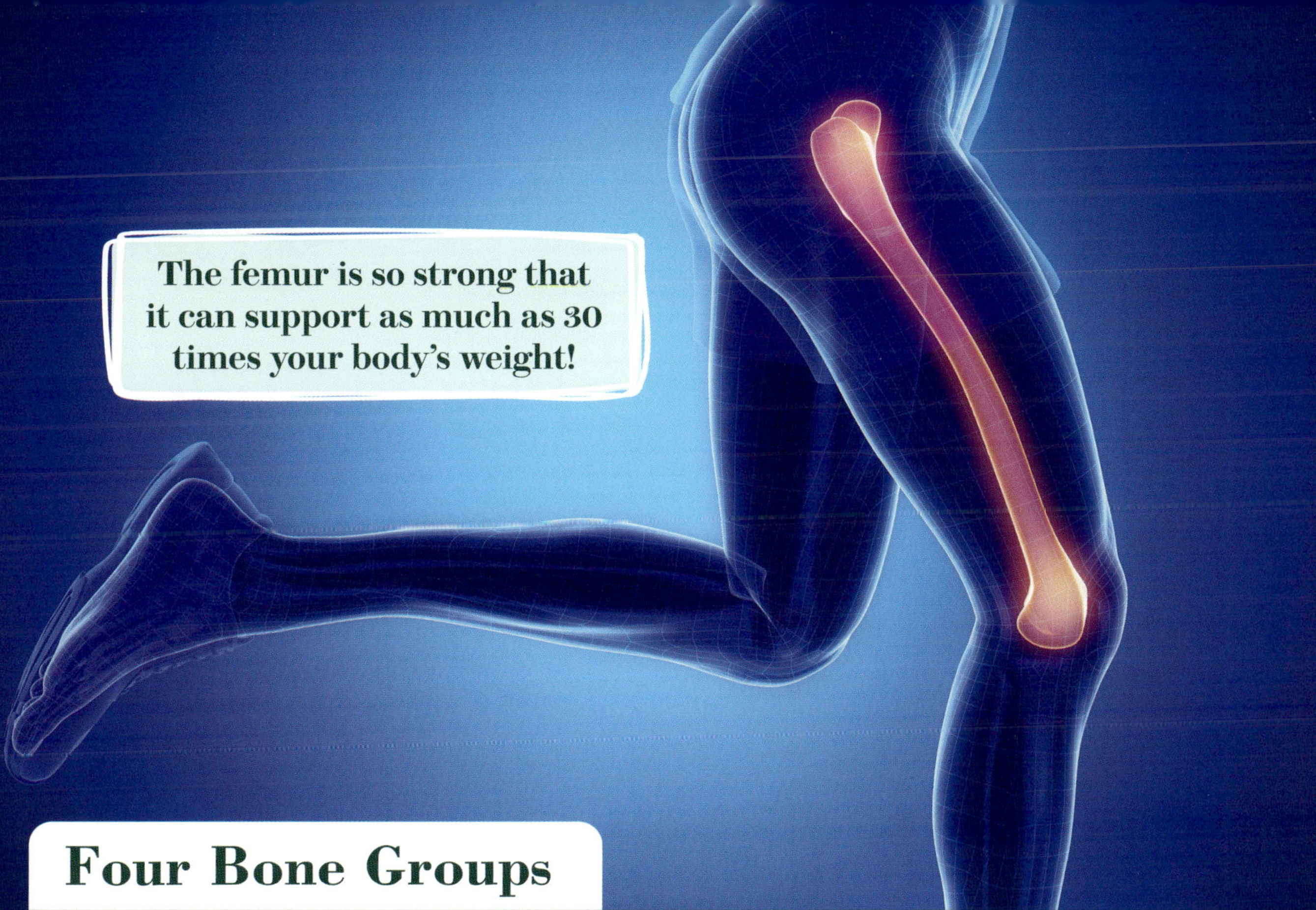

The femur is so strong that it can support as much as 30 times your body's weight!

Four Bone Groups

Bones are placed into four groups based on their shape. Long bones, such as the femur in the leg, support weight or act as a **lever**. Short bones, such as the carpals of the hand, provide support but little movement. Flat bones, such as the scapula (shoulder blade), protect organs and provide a large area for muscles to attach to. Irregular bones have odd shapes, like the vertebrae, or bony segments of the spine.

ALL ABOUT BONES

The two main groups of bones in the skeletal system are the axial skeleton and the appendicular skeleton. The axial skeleton includes the skull, spine, ribs, and sternum (breastbone). These bones form the axis, or central supportive structure, of the body.

The skull is made up of 22 bones. The eight flat bones of the **cranium** fit together like puzzle pieces. The 14 facial bones attach tightly to the cranial bones. Teeth are rooted in the two jaw bones. The six ear bones, called ossicles, are within the skull, although they aren't part of it. They vibrate when sound waves hit the eardrum and send electrical signals to the brain to allow us to hear.

The spine supports the skull and ribs. The spine also protects the spinal cord, a thick rope of nerve tissue extending from the brain.

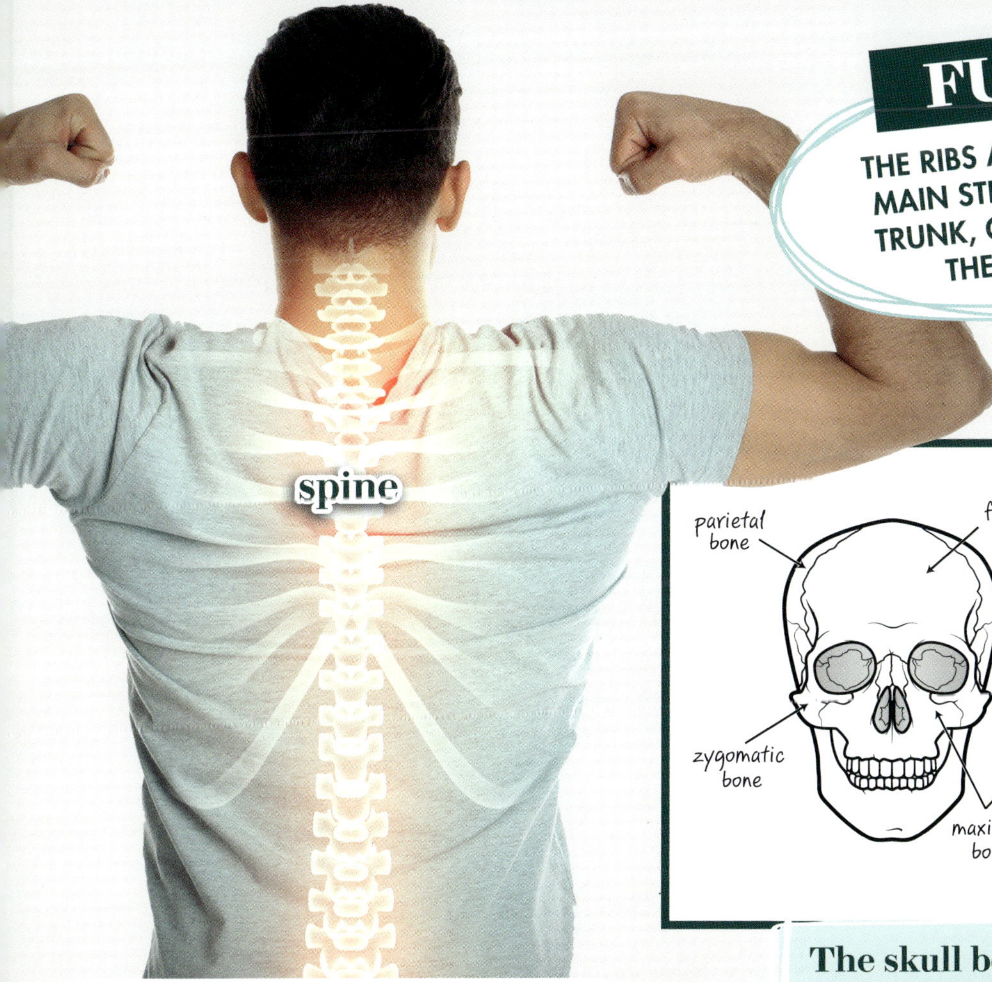

FUN FACT

THE RIBS AND STERNUM FORM THE MAIN STRUCTURE OF THE HUMAN TRUNK, OR TORSO, AND PROTECT THE HEART AND LUNGS.

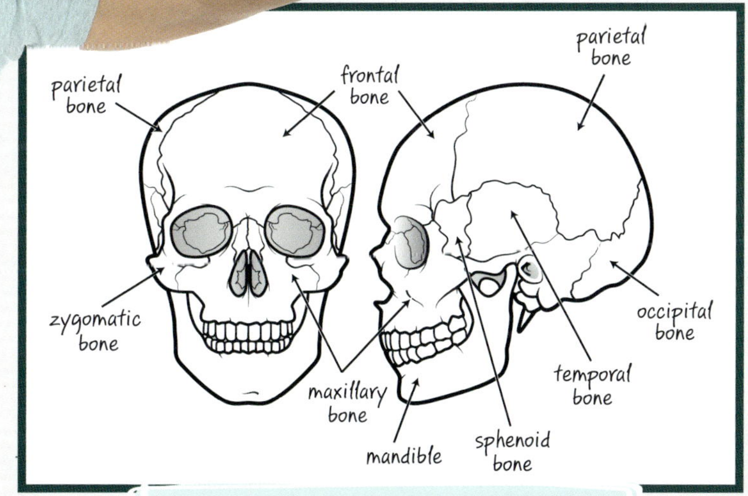

The skull bones protect the brain and sense organs.

Explore the Spine

A baby's spine is made up of a chain of 33 irregular bones called vertebrae. They surround the spinal cord somewhat like beads on a string. Many bones and muscles connect to the vertebrae. The vertebrae form curves as they go down. The top vertebra, the atlas, supports the skull. The bottom section of the spine—the coccyx or tailbone—is made up of three to five increasingly smaller vertebrae that become fused together in adults.

The appendicular skeleton is composed of two main sections that attach to the spinal column by way of "girdles." The pectoral girdle is made up of two clavicles (collarbones) and two scapulas (shoulder blades). It supports the bones of the arms and hands. The pelvic girdle is the pelvis.

The bones of the arms and legs are similar to each other. Ball-and-socket joints connect the humerus (upper arm bone) to the scapula and the femur (thighbone) to the pelvis. The humerus and femur are two of the strongest bones in the body. The radius and ulna run from the elbow to the hand. The tibia and fibula run from the knee to the foot. Most fingers and toes (other than the big toe and thumb) have three phalanges, or bones, that allow movement.

FUN FACT

THE PELVIS IS IMPORTANT FOR SITTING AND STANDING. WHEN YOU SIT, IT SUPPORTS THE UPPER BODY'S WEIGHT. WHEN YOU STAND, IT TRANSFERS THAT WEIGHT TO THE LOWER BODY.

The Skeletal System

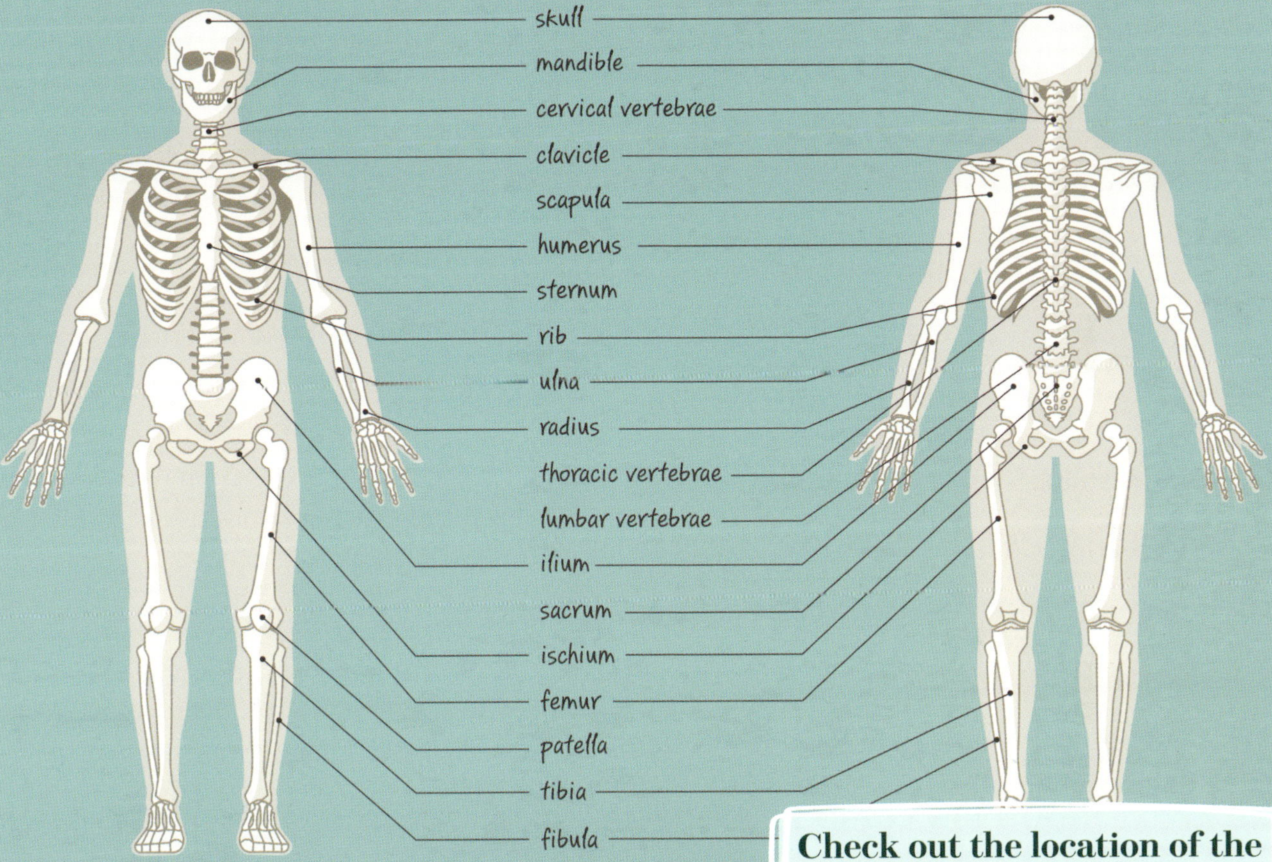

Check out the location of the major bones of the skeletal system with this diagram.

Short Bones

Where they meet the long bones above them, the hands and feet have rows of short bones that slide over each other to allow movement. In the hands, these are called carpals and metacarpals. In the feet, they're called tarsals and metatarsals. Each foot also has a heel bone. The jointed bones in our fingers and toes are called phalanges. All of these bones work well together to help us move in many ways.

WHAT ARE BONES MADE OF?

You might think bones are just one solid object. However, there's a lot inside of them! The outer layer, called the cortex, is made out of compact bone, which is smooth and very hard. The tissue in compact bone is pressed tightly together, which is where it gets its name. Narrow channels carrying nerves and blood vessels run through compact bone. Bones are also surrounded by a thin layer of tissue called periosteum, which contains nerves and blood vessels.

Inside the compact bone is cancellous bone. It's made up of a network of crisscrossing pieces of bone called trabeculae (truh-BEH-kyuh-lee). This network may look **fragile**, but it's actually very strong. The center of many bones contains marrow. This spongy substance helps with blood cell production.

FUN FACT

THE TRABECULAE LINE UP TO GIVE THE BONE THE BEST SUPPORT, KIND OF LIKE THE RAFTERS INSIDE A ROOF.

What we see when we look at a skeleton is the compact bone, also called cortical bone.

Four Kinds of Bone Cells

osteoblasts
make new bone and help repair damaged bones

osteocytes
take in nutrients from blood vessels and get rid of waste

osteoclasts
break down old bone tissue

osteogenic cells
become osteoblasts when needed

Parts of a Bone

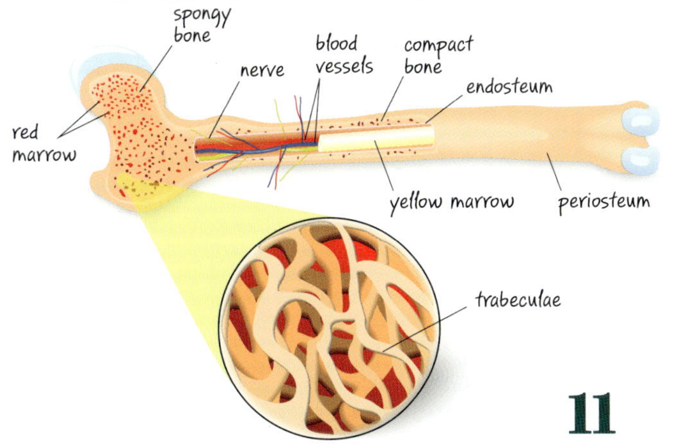

JOINTS AND MOVEMENT

How do bones move? Movement happens at joints, or the areas where two or more bones meet. Most joints are synovial (suh-NOH-vee-uhl). They allow us to move our body and absorb shocks. A synovial joint, such as the knee, is surrounded by a sac called the joint **capsule**. It's lined with a special tissue that makes synovial fluid, which acts as a **lubricant** to allow the joints to bend and flex easily. A special type of **cartilage** connects and cushions the bones that meet inside the capsule.

Another kind of joint, such as those in the pelvis and between the vertebrae of the spine, allows limited movement. The joints of the skull are a third kind, which don't move but allow for growth in childhood.

FUN FACT

AT BIRTH, THE JOINTS OF THE SKULL ARE SOFT AND ALLOW FOR GROWTH. BABIES HAVE TWO "SOFT SPOTS" ON THE SKULL CALLED FONTANELS.

Ligaments hold bones together around the joints. Tendons fasten muscles to bone. These strong, flexible cords allow you to run and jump!

Synovial Joints

The human skeleton has four main kinds of synovial joints. Elbows and knees are hinge joints. They allow back-and-forth movement in one direction. Ball-and-socket joints like the hip and shoulder allow for the most movement. Saddle joints, such those found in the thumbs, allow back-and-forth and up-and-down movement. Plane joints, such as those between the metacarpal bones in the hand, allow limited sliding motions.

YOUR BODY'S SUPPORT SYSTEM

The skeleton is a creature's support system. The human skeleton is called an endoskeleton because it's inside the body. Some animals, such as insects and crabs, have exoskeletons, which are on the outside of their bodies.

Imagine your body without your skeleton. You'd be a pile of jelly. That's because your skeletal system is like a frame for your body. The body's muscles and organs are anchored to the skeletal system, providing them with support. This holds the body together and gives it shape.

The bones and joints work with our muscles and allow us to move in many different ways. Many bones act as levers to help move objects. Our bones allow us to write and speak. They also allow us to play sports, swim, and run around.

The spine is the body's main support structure. All the bones connect to it directly or through other bones, and it allows us to sit and stand upright.

BONES AS PROTECTION

Without bones, your tissues and organs would be exposed to the world and easily hurt. One of the most important organs to protect is the brain. Luckily, the bones of the skull form a shell around the brain and top of the spinal column. They also give some protection to the eyes, nose, and mouth. They're like a helmet!

The lungs, heart, and other organs are protected by the sternum and the ribs. Together, these bones form the rib cage. The sternum is a thick, T-shaped bone in the center of the chest. The ribs begin at the spine and curve around toward the sternum. Cartilage joins all but the bottom two ribs on each side to the sternum. The cartilage is flexible, allowing the rib cage to move when we breathe.

FUN FACT

THE VERTEBRAE OF THE SPINE PROTECT THE SPINAL CORD. THE VERTEBRAE ARE CONNECTED BY ROUND PIECES OF CARTILAGE COMMONLY CALLED DISCS, WHICH ALLOW THEM TO FLEX AND MOVE.

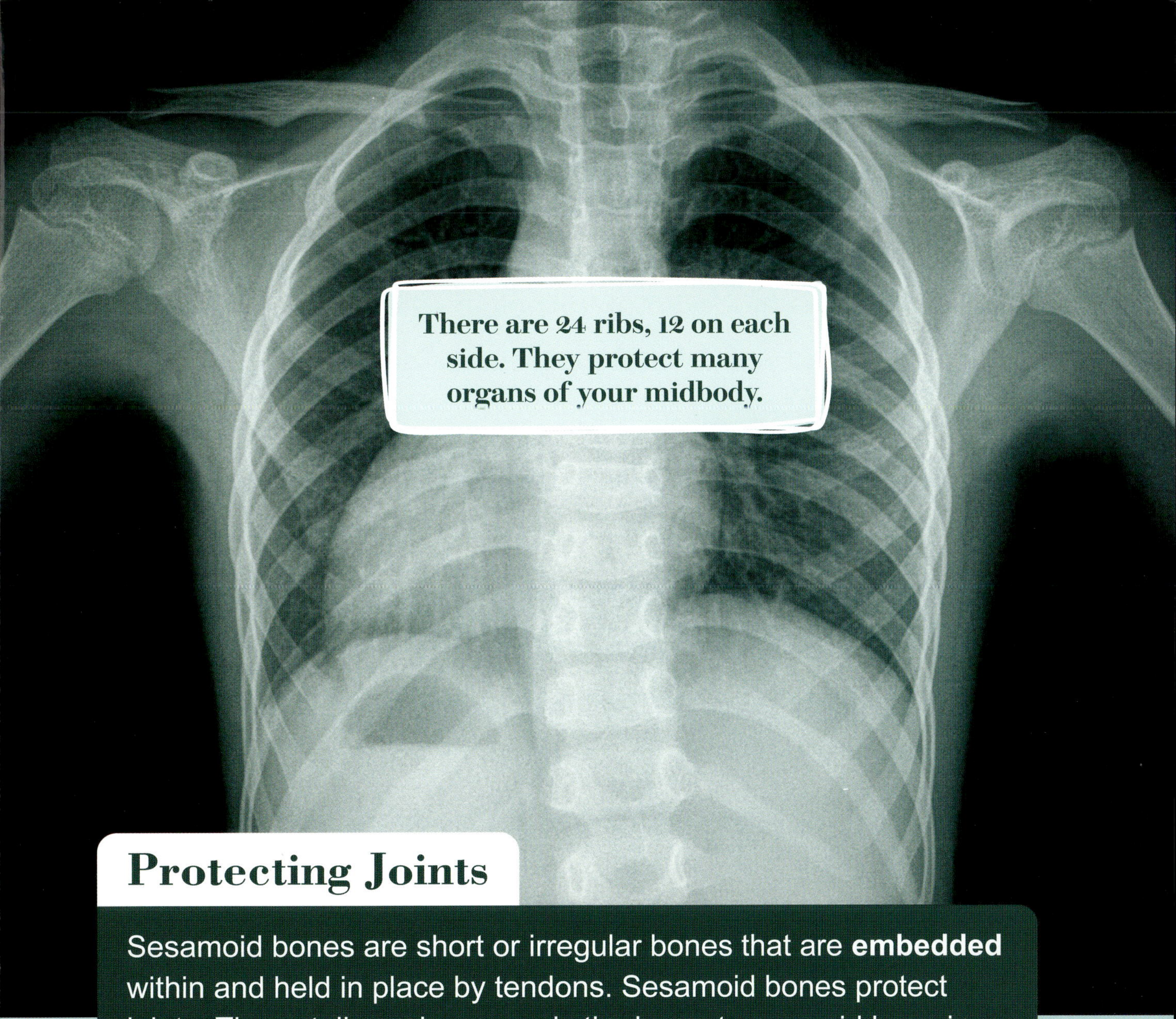

There are 24 ribs, 12 on each side. They protect many organs of your midbody.

Protecting Joints

Sesamoid bones are short or irregular bones that are **embedded** within and held in place by tendons. Sesamoid bones protect joints. The patella, or kneecap, is the largest sesamoid bone in the body. It protects the synovial joint of the knee. The patellar tendon keeps it in place. Other sesamoid bones can be found in the metacarpals of the hands and the metatarsals of the feet.

CELL PRODUCTION

Your body is made up of trillions of cells, and bones play an important role in cell production. They're also an important storehouse for minerals.

Human bones are blood factories! The center of many bones contains a jelly-like substance called bone marrow. There are two types of bone marrow—yellow and red. Fat is stored in yellow marrow. Red marrow creates blood cells. Red blood cells carry oxygen to organs throughout the body. White blood cells help fight illnesses. Along with the liver and spleen, red marrow also helps destroy old red blood cells. From birth to about age seven, all marrow is red. In adults, red marrow is found in the vertebrae, hips, rib cage, skull, and at the ends of the long bones of the arms and legs.

FUN FACT

THE BODY CAN CHANGE YELLOW MARROW INTO RED MARROW UNDER CERTAIN CONDITIONS—FOR EXAMPLE, IN CASE OF SERIOUS BLOOD LOSS.

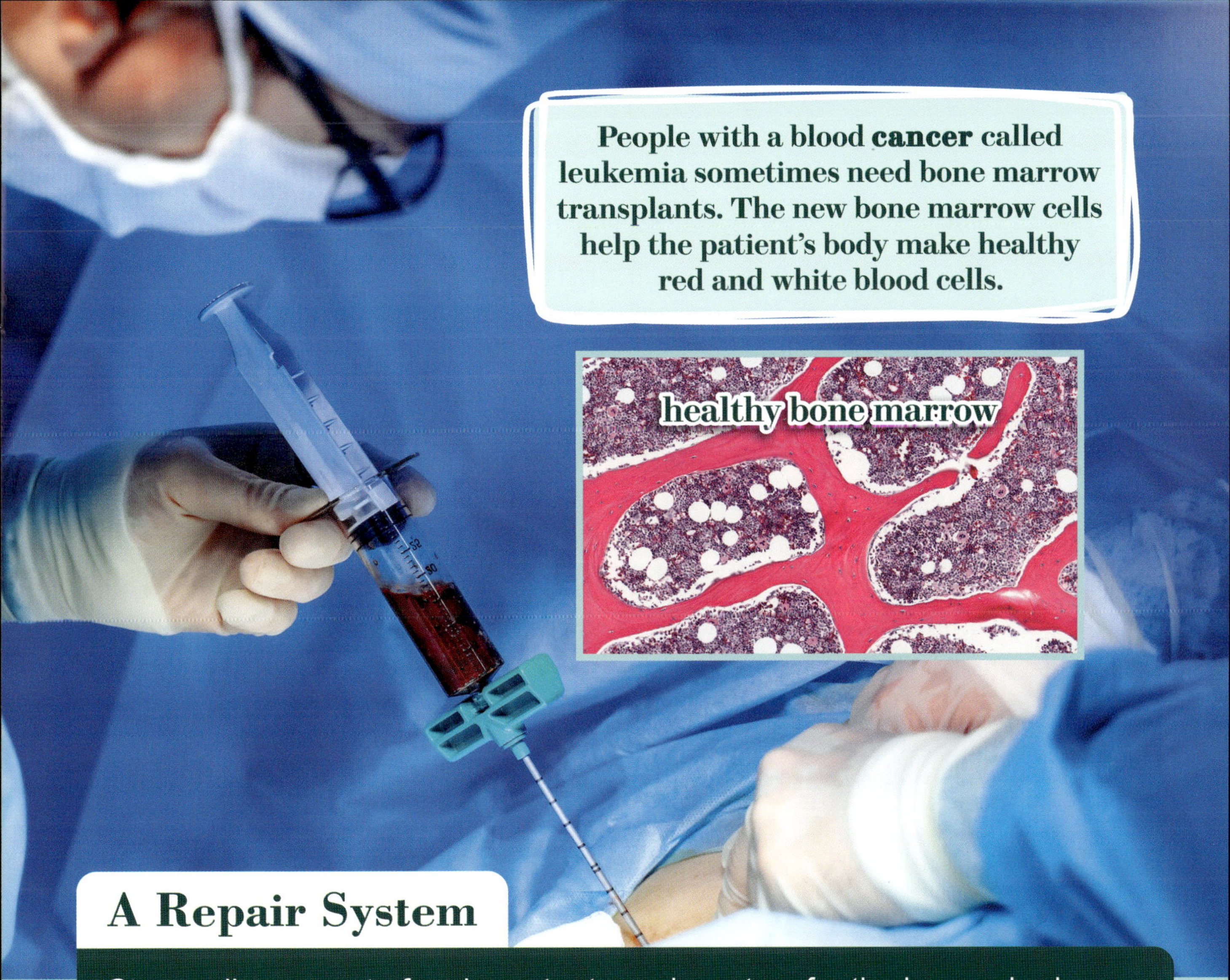

People with a blood cancer called leukemia sometimes need bone marrow transplants. The new bone marrow cells help the patient's body make healthy red and white blood cells.

A Repair System

Stem cells are part of an important repair system for the human body. They're capable of replacing damaged or worn-out cells. Scientists study stem cells to find out how the body repairs itself. Bone marrow contains stem cells. These cells produce new cells by dividing. Some new cells continue as stem cells. Other new cells become different types of blood cells. This is how the bones replace old blood cells.

Blood flows through blood vessels, which are like highways. Blood vessels are found throughout the bones. Blood cells made by marrow pass through the blood vessels and into the rest of the body. The blood vessels also allow the body to send excess minerals to the bones for later use.

The two most common minerals in the body are calcium and phosphorus. They're important to the way the body functions. Cells use these minerals during chemical reactions. When there's too much calcium or phosphorus in the bloodstream, the excess is stored in the bones for later use. When there's not enough of these minerals in the bloodstream, they're released by the bones.

Dairy foods—such as milk and cheese—are a good source of calcium. Beef, nuts, seeds, and chocolate contain phosphorus.

The minerals calcium and phosphorous also help keep your bones and teeth strong.

Bones Make Hormones!

Scientists have discovered that the bones are part of the endocrine system. This system is made up of organs called glands that make chemicals called hormones. Hormones are chemicals that affect the way our bodies function. Bones make the hormone osteocalcin, which helps make our bones strong. Osteocalcin also helps control the body's use of sugar and controls where fat is stored in the body. Scientists think osteocalcin may be used to fight **diabetes** in the future.

SAME, BUT DIFFERENT

Most humans have a similar skeleton. However, there are differences between people born **biologically** male and those born biologically female. Generally, male bones are heavier and longer. Female bones finish developing around age 18. Male bones finish developing around age 21.

Male and female skulls look different. Male skulls have larger, stronger features. Brow ridges, for example, are generally larger in male skulls. The biggest differences between male and female skeletons can be seen in the pelvis. The largest parts of the pelvis, called the ilia, are wider in female skeletons. The space between the two halves of the pelvis is wider in women than in men. The tailbone is more moveable in female skeletons, too. This makes giving birth possible for some women.

FUN FACT

MALE CHINS ARE USUALLY SQUARER. FEMALE CHINS ARE OFTEN MORE ROUNDED AND NARROWER.

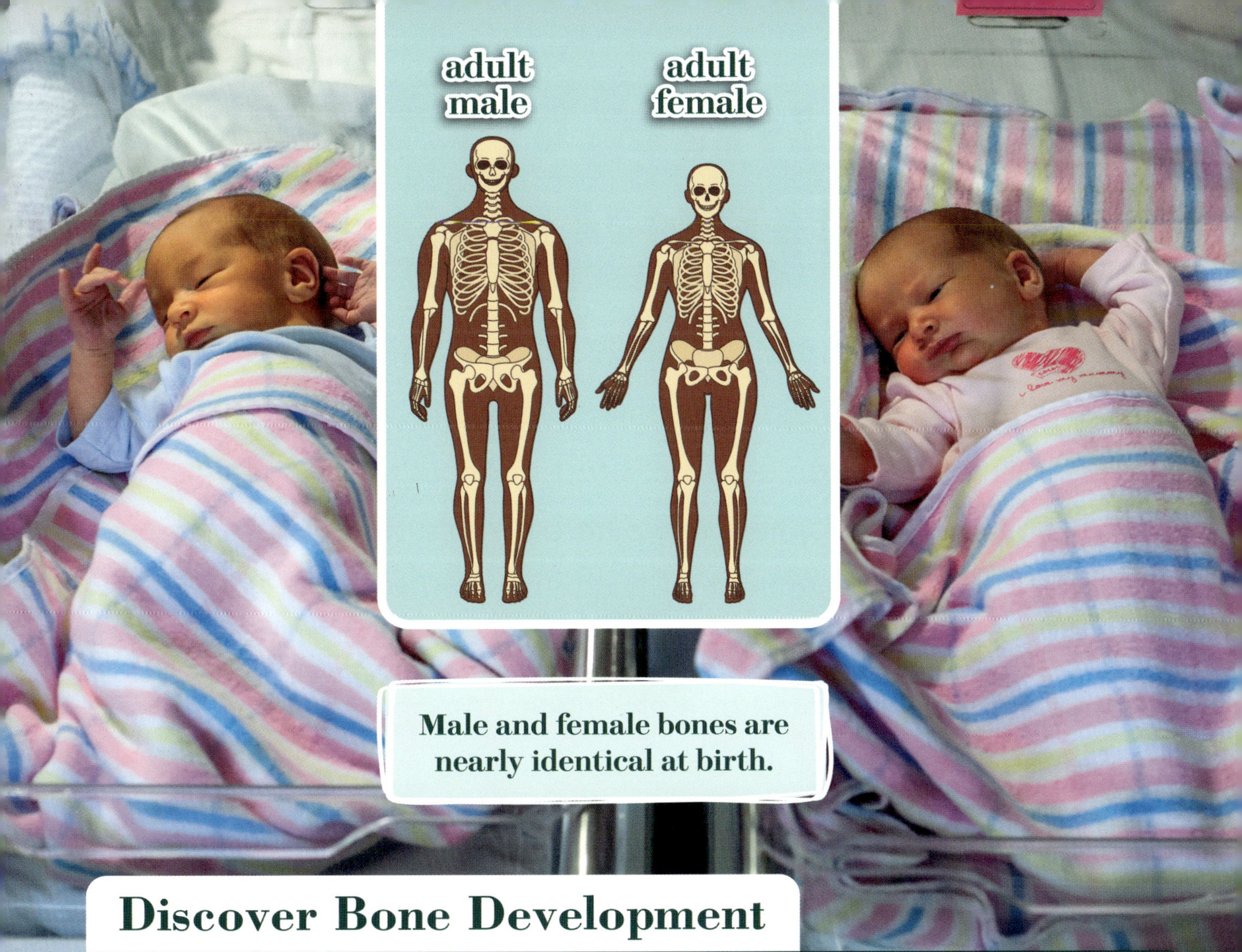

Male and female bones are nearly identical at birth.

Discover Bone Development

A newborn's bones are soft and flexible. Most are very similar to cartilage. Even the bones of the skull are soft. As a child grows, its bones harden. In women, the hormone estrogen causes the bones to finish growing earlier than in men. In men, the hormone testosterone gives the bones more time to develop, allowing them to achieve greater length and mass. The formation of bone tissue is called ossification.

BREAKS AND BRUISES

Bones withstand a lot of pressure in your everyday activities. They can handle some bumps and falls. However, sometimes, they can bruise or break if enough pressure is placed on them. There are different kinds of fractures, or breaks. A hairline fracture is a minor break that's often missed on an X-ray. A comminuted fracture occurs when a bone breaks into three or more pieces. An open fracture occurs when part of the bone pierces the skin. Children's bones are more flexible than adult bones. Because of this, children often experience "greenstick" fractures. This occurs when a bone bends and splinters instead of breaking.

People with fractures may need to wear a brace or cast. This allows the bone time to heal and keeps it from moving. Serious injuries may require surgery.

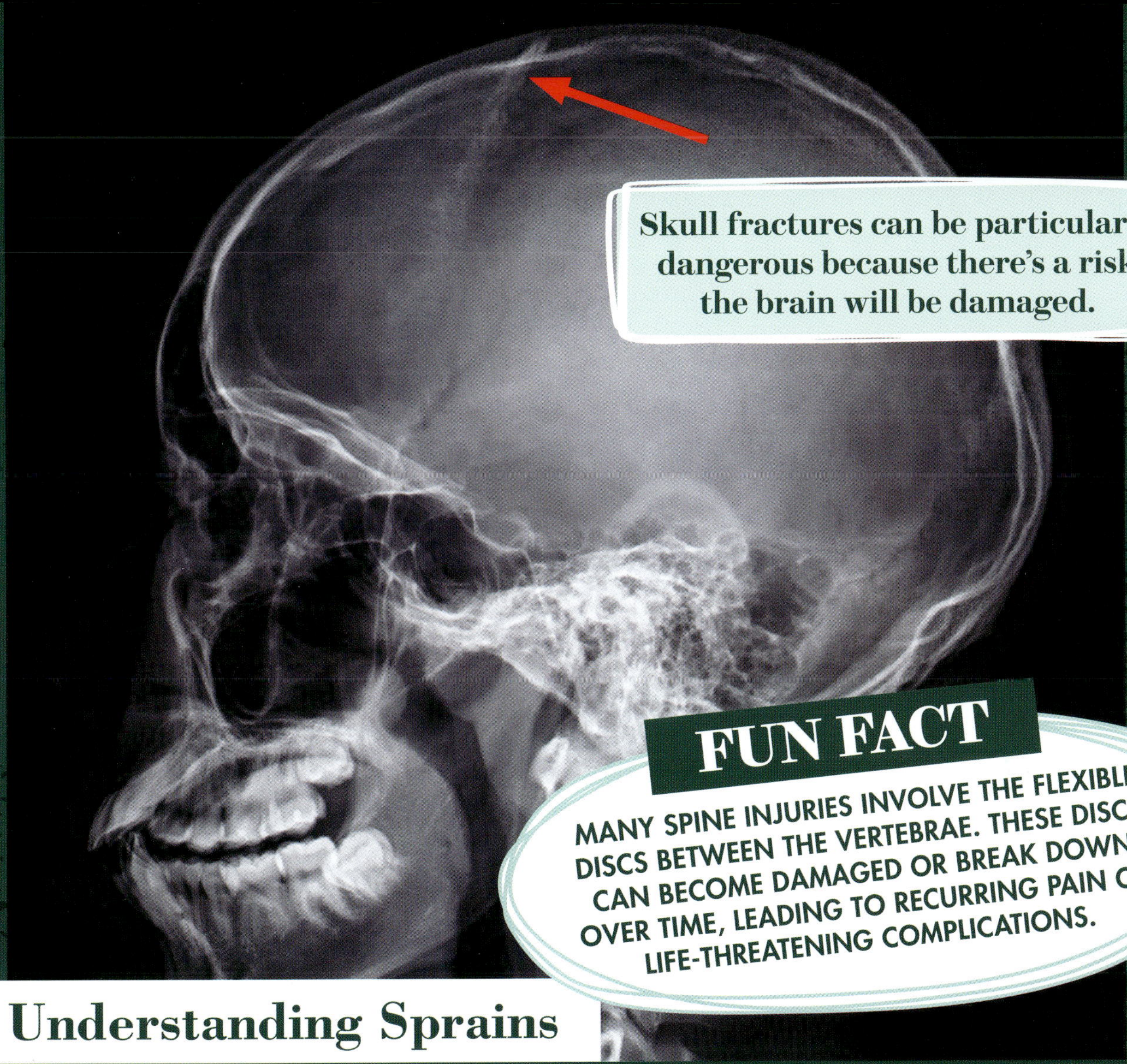

Skull fractures can be particularly dangerous because there's a risk the brain will be damaged.

FUN FACT

MANY SPINE INJURIES INVOLVE THE FLEXIBLE DISCS BETWEEN THE VERTEBRAE. THESE DISCS CAN BECOME DAMAGED OR BREAK DOWN OVER TIME, LEADING TO RECURRING PAIN OR LIFE-THREATENING COMPLICATIONS.

Understanding Sprains

A sprain is an injury to the ligament around a joint. It can happen when a ligament is stretched too far or when it tears. Many athletes sprain their ankles. A dislocation occurs when the ends of two bones are forced apart. This can cause damage to ligaments, nerves, and blood vessels. Sprains and dislocations cause pain and swelling. Minor joint injuries heal with rest, but major injuries sometimes require surgery.

BONE DISEASES AND DISORDERS

Some people suffer from diseases of the bones, including cancer. Osteoporosis is the most common bone disease. This illness is marked by a loss of bone **density** over time. Osteoporosis occurs when the body loses too much bone tissue or doesn't make enough. It results in brittle, weak bones. Paget's disease causes unnatural growth and weakness of the bones. It affects about 1 percent of adults in the United States.

Spina bifida is a **birth defect** that occurs when an unborn baby's spine doesn't form properly. The condition may be discovered during pregnancy or soon after birth, and is often treated with surgery. While some people with spina bifida use crutches or a wheelchair, others are able to walk on their own.

Osteoporosis can be prevented by increasing the amount of calcium in the diet.

FUN FACT

OSTEOPOROSIS AFFECTS WOMEN MORE THAN MEN—ABOUT ONE IN FIVE WOMEN OVER THE AGE OF 50 IN THE UNITED STATES HAS OSTEOPOROSIS.

Joint Pain

Arthritis is painful swelling in one or more joints. Some kinds are caused by injuries or general usage over a long time. Other kinds are caused by infections or problems with the **immune system**. The most common form is called osteoarthritis. It's marked by the breakdown of cartilage between bones, which causes the bones to grind together. The joints become stiff, swollen, and painful. Some drugs help reduce pain, but there's no cure for arthritis.

KEEP YOUR BONES HEALTHY!

Your bones protect many of your body parts, so it's important to protect your bones! Make sure you get enough calcium each day, through food sources like milk, yogurt, and almonds. Strengthen your muscles and maintain healthy bones by doing strength exercises, such as lifting weights.

You can protect your bones from injury by staying safe when playing sports or doing risky activities. Wear a helmet while doing activities such as biking, skateboarding, and riding a scooter. Wear all the protective gear that's recommended for certain contact sports, such as football. Wear a mouth guard to protect your teeth. Injuries when you're young can affect you as you get older. When you take care of your bones, they can be the support system your body needs!

Wear pads on your knees and elbows during activities like roller-skating.

FUN FACT

ADULTS REACH PEAK BONE MASS IN THEIR EARLY TWENTIES. AS WE GROW OLDER, WE CAN LOSE BONE DENSITY.

GLOSSARY

biologically: Having to do with one's biology, or the physical features one is born with.

birth defect: A problem with how the body works or is formed that is present from birth.

cancer: A disease caused by the uncontrolled growth of cells in the body.

capsule: A sac enclosing a body part.

cartilage: Tough, flexible tissue in the body.

cranium: The part of the skull that covers the brain.

density: The amount of a material in a given area.

diabetes: A disorder that causes the body to produce excess urine and causes high levels of sugar in the blood.

embedded: Set solidly within something else.

fragile: Easily breakable.

immune system: The parts of the body that fight germs and keep it healthy.

lever: A tool used to lift a load.

lubricant: A substance that allows two surfaces to slide easily over each other, such as grease.

FOR MORE INFORMATION

Books

Cernak, Linda. *The Human Body Encyclopedia.* Minneapolis, MN: Abdo Reference, 2023.

Edwards, Jonas. *The Skeletal System.* New York, NY: Gareth Stevens Publishing, 2022.

Vonder Brink, Tracy. *Skeletal System.* St. Catharines, ON: Crabtree Publishing, 2023.

Websites

Bones and the Human Skeleton
www.ducksters.com/science/bones.php
Explore more fun facts about the human skeleton.

Quiz: Bones
kidshealth.org/en/kids/ssquiz.html
Test your knowledge about bones with this fun quiz.

Skeletal System
kids.britannica.com/kids/article/skeletal-system/353778
Learn about the skeletal system's jobs and how it's connected to the rest of your body.

Publisher's note to educators and parents: Our editors have carefully reviewed these websites to ensure that they are suitable for students. Many websites change frequently, however, and we cannot guarantee that a site's future contents will continue to meet our high standards of quality and educational value. Be advised that students should be closely supervised whenever they access the internet.

INDEX

adults, 4, 7, 18, 23, 24, 26, 29

appendicular skeleton, 6, 8

axial skeleton, 6

babies, 4, 7, 12, 26

blood vessels, 10, 11, 20, 25

bone cells, 11

bone marrow, 10, 18, 19, 20

brain, 6, 7, 16, 25

cartilage, 12, 16, 23, 27

cell production, 10, 18

endocrine system, 21

exoskeletons, 14

femur, 4, 5, 8

fractures, 24, 25

girdles, 8

joints, 8, 12, 13, 14, 17, 25, 27

organs, 4, 5, 7, 14, 16, 17, 18, 21

pelvis, 8, 12, 22

skull, 6, 7, 12, 16, 18, 22, 23, 25

spine, 6, 7, 12, 15, 16, 25, 26

sprains, 25

sternum, 6, 7, 16

structure, 4, 6, 7, 15

vertebrae, 5, 7, 12, 16, 18, 25